DEMOLITION EQUIPMENT

Hans Halberstadt

Motorbooks International
Publishers & Wholesalers ®

To Chris and Stephanie Hope

First published in 1996 by Motorbooks International Publishers & Wholesalers, PO Box 1,729 Prospect Avenue, Osceola, WI 54020-0001 USA

Motorbooks International books are also available at discounts in bulk quantity for industrial or sales-promotional use. For details write to Special Sales Manager at the Publisher's address

Library of Congress Cataloging-in-Publication Data Available

ISBN 0-7603-0042-9

On the front cover: A Komatsu PC220LC excavator tears a reinforced concrete building into rubble.

On the frontispiece: The Pacific Palisades Hotel on its way from twenty-two stories to none.

On the title page: Two big excavators, a top-of-the-line John Deere 992 and Hitachi 270, dig into the rubble of the newly dropped Pacific Palisades Hotel.

On the back cover: A dangerous time for the operator of a Cat 231D LC. Every bite of the bucket creates tremors throughout the remains of the building, threatening an uncontrolled collapse.

Printed in Hong Kong

Contents

Acknowledgments

It would be virtually impossible to produce these kinds of books without the generous support of people in the industry who explain how things work and occasionally allow inquisitive authors (particularly me) play with the machines.

Motorbooks titles about Caterpillar products, in particular, have been blessed with the cooperation of the place where it all began, Holt Brothers of Stockton, California; Jim Hasty did the honors this time, providing lots of literature and instruction on Cat products. If you're in the market for a Cat product, give the nation's number one heavy equipment dealer a call.

Mike Taylor, Executive Director of the National Association of Demolition Contractors (NADC), also provided lots of help—despite an industry tradition of shyness.

Lots of other folks contributed to this book:

- Corry Goumans of Pacific Blasting
- Stan Holtby of CDI
- Mike Morrison and Bill Moore of Brandenberg Demolition
- Mike Murphy of Komatsu
- Tom Robinette of Robinette Demolition
- Dave Rogers, the J. I. Case Company's friendly photographer
- Jim and Jarred Redyke of Dykon
- Jack Stewart of Allied-Gator

Thanks to all!

Introduction

Until about five o'clock in the afternoon of October 17, 1989, I actually enjoyed the earthquakes that we get here in California every few years; the windows of our house rattled, the ground might vibrate a bit, or perhaps the electricity would go off for an hour or two. No big deal. But that afternoon, just before the Giants took the field at Candlestick Park for the World Series, was different.

I was upstairs in our 1908 bungalow-style frame house. The earthquake began in the usual way, with a quiet rumble and a rattle from the windows. But this time the intensity was just a little stronger, the rattle a little louder, and it didn't stop for what seemed like a long time. I could hear glass breaking downstairs and the thud of household objects hitting the floor. I thought to myself, as the shaking continued, *people are getting hurt out there.* Then it stopped.

The power was out. I went downstairs; there was broken glass everywhere, but the house was still standing. Outside, my neighbors emerged to compare notes; no real damage seemed to have been done, nobody seemed injured...although we could hear sirens wailing in the distance.

A battery-powered radio tuned to a San Francisco news station began to provide more information, and it quickly became apparent that people had indeed been getting hurt. Apartment buildings in the Marina district of San Francisco had collapsed; others were on fire. The Cypress Avenue portion of Interstate 880 through Oakland collapsed at the height of the rush hour, the radio reported, with the probable loss of hundreds of lives. The Bay Bridge had collapsed, according to another story. Reports from outlying communities told of widespread devastation. Santa Cruz,

This mess is the partial remains of part of Interstate 880 known as the Cypress structure. Located in the city of Oakland, California, the Cypress structure was destroyed in the 1989 Loma Prieta earthquake. It is slowly being hammered into rubble by the impact of a 5,000lb cast iron ball swung by a crane.

about seventy miles south of San Francisco, lost much of its charming old brick downtown, including a favorite of mine, the old Cooper House, a classic nineteenth century courthouse with a vast stained glass skylight. There was, the radio reported, death and destruction everywhere.

Well, it turned out to be less than Armageddon; lives were lost, but not nearly as many as the media hype had indicated. Several freeways had collapsed and others were damaged, and Santa Cruz's main street was in fact full of brick rubble. The power was out for about eighteen hours—and all of my neighbors gathered for our best-ever block party. While none of my friends or family was injured, some of our houses were damaged and one friend's house was a total loss. We all were reminded just how fragile our homes and public buildings really are. And I don't really enjoy earthquakes any more.

one

Wrecking Industry Primer

The San Francisco earthquake of 1989 twisted and snapped bridges, knocked down houses, and generally wreaked havoc with the city, creating a lot of work for wrecking companies. Hundreds of buildings were quickly determined to be total losses. They had to be cleared, immediately, because of the hazard they created for public health and safety. Streets were blocked, freeways had to be removed, and lots were cleared so replacement structures could be built.

While the dust began to settle, a huge effort by an unusual industry began. Contractors and specialists of many kinds began the task of clearing away the rubble of shattered buildings and fallen freeways. While some of the bulldozers and wheel-loaders were owned and operated by building construction companies, the most important jobs were performed by demolition specialists—people from the "wrecking" industry with special equipment, skills, and experience.

The building demolition industry's work was, for a while, in a bright and friendly spotlight. That was an unusual experience for this small industry, which is usually harangued for pulling down some landmark or other. But the earthquake revealed that many of the structures in our communities are really quite fragile. We thought the freeway was sound, and it wasn't; we thought the old brick buildings in downtown Santa Cruz were quaint, and they were actually death traps.

Thousands of buildings in the United States and Canada are pulled down every year, for many reasons. Some are damaged by earthquakes, floods, fires, tornadoes, hurricanes, decay, accidents, or when sink holes develop beneath them. Others have outlived their utility and are removed by their owners for replacement structures. Still others (like the beloved old Cooper House) look perfectly sound yet are ready to collapse; some of these are pulled down for safety reasons.

The National Association of Demolition Contractors (NADC)

The people who specialize in demolition belong to a small, interesting industry. There are only about 400 companies in the United States that specialize in building demolition. Most are family operations, and none are publicly held corporations. Their trade group is the National Association of Demolition Contractors (NADC). The people who make up this group turn out to be an eclectic bunch.

Robinette Demolition is a good example. Tom Robinette started in the business back in the 1960s as a teenager, learned to run the old mechanical dozers, cranes, and loaders, and started his own company in the mid-1970s after learning the ropes. The company developed a good reputation, a ten-million-dollar line of insurance, and a small, hard-core group of about 100 experienced people. That makes the company one of the largest demolition outfits; there are a few larger in the NADC.

Robinette contracts to demolish the whole building or just to strip out the interior. Robinette Demolition, like most of their competitors, is heavily involved in recycling. Unlike some of the competition, though, they are a union shop, have their own environmental services office, and guarantee legal disposal of debris. While that sort of record is pretty characteristic of members of the NADC, it certainly isn't

LEFT
Boy, what a mess! And the folks who made it are going to have to clean it up, too. That's the carcass of a food processing plant in the background, rapidly being reduced to rubble.

9

Kids, don't try this at home! A Caterpillar 231 excavator takes another bite out of an old brick building in Racine, Wisconsin. These handsome old brick buildings are not reinforced and can be extremely fragile. This one literally fell apart, with large sections of brick wall collapsing after a little prod from the bucket.

true of many smaller freelance operations that sometimes take on demolition jobs.

At last count there were exactly 4,004 companies listed in the United States and Canada who claim to be qualified to take on demolition projects, but the NADC's director, Mike Taylor, estimates that no more than 800 do more than an occasional job. NADC membership is only about 650 and that includes memberships from the manufacturing and support communities, with only about 400 hard-core wrecking companies in North America.

The NADC, despite its small size, has a lot of clout. Trade organizations like the NADC work closely with the federal government to design regulations for the safe handling of hazardous materials such as lead and asbestos as well as safety standards for the use of machines and materials on the job.

The Right Tool for the Job

Demolition projects involve structures built as long ago as 1850 and as recently, at times, as a year or two ago. That means that the kind of materials, the type of construction, and the way the building is stitched together can be radically different. Old nineteenth century brick and wood warehouses look charming but are often extremely fragile and near collapse. Wood-frame houses and structures require completely different tools and techniques for demolition. Formed concrete grain silos from the 1920s—ten stories high, loaded with rebar, and tough as nails—present an entirely different challenge to the wrecker.

While anybody with an excavator and a contractor's license can try to take on a demo job, building contractors get themselves into trouble every year by taking on projects that are more dangerous than expected. Walls and ceilings collapse

Another member of the team, the track loader, hard at work.

unexpectedly, dumping brick and steel on machines and workers; floors fail, dumping excavators into basements; beams and pipes become huge springs, releasing energy at embarrassing moments. And while everybody likes taking down that first wall of an old building, the last wall can be *scary* when the structural integrity is gone and tons of bricks are wobbling in the wind.

So there are extra hazards in the wrecking business not present in building construction. Those hazards have resulted in steep insurance premiums, tight OSHA standards, regulations from cities and states, and regulations from within the

industry itself that have tended to separate the professionals from the part-timers. Even though excavators cost a quarter-million dollars, insurance is often the most costly part of a job. No wonder there are only a few hundred specialist demolition firms in the whole country.

Beyond the high costs and restrictive regulations, knocking down buildings requires enough skill to qualify as an art. "A good, professional equipment operator needs to have experience on all kinds of buildings, from the 1850s to the present," Tom Robinette says. Tom should know, too, because he's been a wrecker since the 1960s and has been in the seat of all types of machines on all kinds of jobs.

"There has to be a balance between the liability of the drop and the size of the machine—those two considerations have to balance," Tom says. "If you have a three-story *building* and a two-story *machine* you begin to have a problem. You can still do the job, but there is a greater element of risk involved. That's why we will usually have a walk-through evaluation of a job before we even bring in the machines. That's when we decide what kind of machines we'll use on the drop. The operator gets to look at the strengths and weaknesses of the structure, at the proximity of adjacent structures that have to be protected, at the hazards and possibilities of the job. A small contractor may not have a machine perfectly suitable to the drop, but will take it anyway. Then the operator begins to sweat! He has to figure out how to take the building down with what he's got—without hitting the building next door, and without having the structure fall on himself and his machine."

The drop is engineered by the project manager and the equipment operators. Unlike when the building was constructed, the demolition phase is fast, dangerous, and to a certain extent unpredictable. An operator might begin to take out a row of steel columns providing support, expecting the structure to weaken and sag in one way and have the building start to react in an entirely different way because of invisible elements like material failure in beams, walls, or flooring. "You have to have VERY fast reflexes," Robinette explains. "It can start to shift on you, start moving around on you in ways you didn't anticipate. You have to grab, pull, poke, or push. An equipment operator has to be like a rabbit, very alert, always with an eye on everything because things happen so quickly! When you pull a chunk out of a wall you watch for the fracture cracks—that shows you how much of the wall is going to come out. Then, as soon as that section free-falls out of your control, you have to watch how the structure reacts. By pulling out that section of wall, the structure might react by dropping a forty-by-forty foot section of the roof."

The high-speed hydraulics of today's machines accentuate the need for operator alertness. Back in the 1960s, when men like Tom Robinette started, sluggish friction clutches and sloppy mechanical linkages made it take a lot longer for an operator to make a big mistake. Today's hydraulic excavators move fast, with lots of power. One slip can knock over a brick wall, or bring down the entire building prematurely. "The speed of today's machines is terrific," says Tom, "but it can also be a problem for someone who's not alert or in control of the machine."

Making the Hook

"Your eyes never leave the building while you are operating," Tom Robinette says. "As you pull down a brick building, one bite at a time, you don't watch the bricks fall to the ground—you look at the building to see how the remaining structure will react. When you pull out that section of wall you'll see fracture cracks appear along the wall—that tells you how big a chunk is coming down. As soon as that piece begins to free-fall out of your control and you're sure it won't hit you or your machine, you watch the roof and the columns to see how they respond. That bite you just took was a controlled removal of material, but immediately after you run the risk of an uncontrolled drop of material. The roof might start to come down, and it might try to come down on you!"

When unplanned collapses threaten an operator or a machine, the speed and power of the hydraulic excavator can be a literal lifesaver. You can quickly raise the bucket and boom to deflect a falling wall or roof; the machine may be damaged but the operator may be saved. Such heroics aren't needed very often. What *is* needed fairly often is prompt action from the operator when a wall starts to lean or a roof starts to sag in unanticipated ways. That's when the operator quickly "catches" the building with the bucket, pulling or pushing at a critically weak place. But that's not part of the normal routine, either.

Instead, the operator pulls out a section, watches the reaction, then decides where he'll

Reinforced concrete like this freeway is extremely difficult to break up, even with modern equipment. This Cat 235 excavator (a middle-of-the-road machine) is equipped with a special concrete-crusher attachment on the end of the "stick."

make his next "hook." That might be another twenty feet of wall, for example, depending on the operator's comfort level. Then it will be time to "hook" a section of the roof.

"At this point," Tom Robinette explains, "an operator might move in a couple of bays, or maybe punch the basement in. If you don't punch the basement in, you run the risk of having the machine fall into the basement. It happens every year, especially to inexperienced operators."

Another thing to watch out for is the sling-shot effect produced by steel beams and pipes that get bent during the demolition, then release their energy, throwing bricks and debris around the job site. The operator must take care so that nobody

RIGHT
A large old shopping center in Raytown, Missouri, falls prey to a Case 980 excavator and the folks from Spiritas Demolition Company, a St. Louis company typical of many in the National Association of Demolition Contractors (NADC). Spiritas is a family operation that specializes in wrecking jobs all across the Midwest.

14

Now here's a ticklish problem—peeling the brick wall of one building away from another. Bruno Ferrari, with the radio, owns the company that's almost finished demolishing one building…and he and the owner of the one next door want to make sure that only one building is demolished.

on the job is endangered by this hazard, including himself. He also needs to protect the excavator; a severed hydraulic hose will put a machine out of commission for a day or more, a costly experience.

The speed of modern excavators increases production but also increases the risks, especially for the operator who isn't fully alert. "The machine can swing six feet in just a second," Tom says, "especially with these modern machines. They are EXTREMELY quick! That speed is usually a great advantage but if you aren't paying attention and the machine gets ahead of you, the machine can easily sweep out a cinder block wall, a column that

you were depending on to keep the roof up—then you can get a collapse!"

Occasionally Tom's company, Robinette Demo-lition, will run across a building where there's no easy, safe, straightforward way to attack: large, clear-span structures like churches or theaters, that can't be taken down progressively without great risk of sudden col-lapse. When that happens, there's an alternate method that provides a good show and always attracts a good crowd of what Tom calls "wrecking groupies." The method is called *cabling* and is quite simple and effec-tive: wrap a very large steel cable around the structure, attach it to one or more of the most powerful dozers

16

That big clamshell bucket is well-suited for demolition jobs. Bruno Ferrari supervises from on high while the oper-ator carves sections of electrical conduit from the wall.

you can acquire, and pull. The entire building comes down all at once, in a cloud of dust.

Houses

Although not a very profitable kind of demo-lition, single-family residences are a common kind of job for wrecking companies. In my home town of San Jose, California, our city bought up and demol-ished hundreds of houses—an entire neighbor-hood—because of its proximity to the local airport and problems with jet aircraft noise. That neighbor-hood is now open space, a large park in the middle of town. Other houses were demolished because fire or other structural damage made them unsafe

and beyond repair. And yet others were dropped to make way for office buildings, freeways, hospital expansion projects, or other developments.

"The typical house takes less than an hour to drop," Tom Robinette reports. "When I was a kid, start-ing out in this business, I used to race the clock—I got where I could drop a two-story house in twenty-seven minutes!" Tom used an old 977 Cat front-end loader for those house jobs, taking big bites out of the build-ing. That was before the backhoe and grapple, or the tracked front-end loader, the machine of choice today.

Robinette Demolition drops hundreds of houses in the Chicago area each year. And they've learned a few things about dropping them, as Tom

explains: "One thing you always do before dropping a house is to go through it and make sure nobody's living in it! And, too, it isn't unusual to lift the front porch up and have a family of raccoons come out. We've shut down many a job while the animals got out of the way!"

Machine Costs

"We used to be able to buy a machine for a dollar a pound," Tom says, "a 40,000lb machine for $40,000; that same machine will cost you $250,000 today." Well, actually, it can be a lot more than that. A quarter of a million bucks is just for a run-of-the-mill machine; the biggest cost a million dollars or more.

The purchase price is just the beginning. The cost of owning and operating are a lot more than that, particularly in the demolition industry where routine use consists of severe service and rapidly reducing resale value. Interest on the loan, operator pay ($25–30 per hour), fuel (a Cat 953B track loader, for example, will use about 5gal per

The tracked front loader bats clean up on the wrecker's all-star team. This older machine's bucket controls are polished by use and a knob has disappeared, but it scurries around the site with all the agility of the newer machines—at a daily rental rate half that of a fresher loader of the same size.

hour), and repairs (from $5–16 per hour) all add to operation cost of demolition equipment. Demolition equipment owners need to carefully keep track of expenses, or these costs can quickly drive them out of business.

ABOVE
Demolition work is hard on machines and sometimes on people, too. Both get dirty and sometimes a bit worn around the edges. But both tend to be tough, muscular, efficient, and hard working, like this Cat 955K track loader.

RIGHT
Only a few years ago, a building like this might be dropped by hand—men with sledgehammers and pry-bars. The hydraulic excavator changed that, allowing an operator to apply 30,000lb of force with tremendous precision. Here, the operator uses a tooth on the bucket to gently prod a section of wall before pushing it into the rubble pile.

Knocking down the building is the easy part—sorting out the debris and loading it on the trucks to clear the site involves a lot of people and machines. The track loader scurries around in front and alongside the excavator.

RIGHT
Here's an excavator clearing rubble, loading debris trucks with a two cubic yard grapple-type bucket. While track loaders are often used for this work, the excavator can sometimes be more productive; the track loader is faster on its tracks, but the excavator's rapid swing speed can sometimes provide a lower cycle time.

two
Excavators

Excavators are a comparatively new kind of construction vehicle, a design that has revolutionized the construction and demolition industries. While tracked bulldozer-type vehicles have been around for over eighty years, the hydraulic excavator has only been on the job site since the 1960s.

"The predecessor of the hydraulic excavator is the old steam shovel," Komatsu's Mike Murphy explains. "The next generation used internal combustion engines, still with cable drive for the arm and bucket. The hydraulic version started appearing in the late 1960s. In the thirty years the design has been around, people have discovered the advantages of this unique tool."

The concept is based on a huge, extremely powerful arm that is quite similar to the human arm, except it can reach over 40ft and can crush concrete with its grip. An excavator's arm is also incredibly precise; while you can't quite perform brain surgery with one, a skilled operator can manipulate the attachment on the end of the "dipper-stick" with the speed and dexterity of a hand-held tool. This precision is the result of sophisticated, computer-controlled hydraulics.

Excavators are powered by relatively small engines, especially when compared to other machines in the demolition industry. A 40,000lb—heavy equipment is rated by its weight—excavator will be powered by a 120hp diesel engine; the engine in a wheel loader or dozer in that same 40,000lb class will have about twice as much horsepower. Why the big difference? Well, excavators are relatively static machines; they don't move around a job site nearly as much as a dozer or a wheel loader, and when they do move it tends to be a pivot around a ball bearing. The thing that moves the most on an excavator is the arm, not the tracks, and that permits a smaller power plant, with less fuel consumption—and bigger potential profits.

"An excavator is a very efficient machine," says Komatsu's Mike Murphy. "If you can do a job with an excavator instead of a wheel loader or dozer, it will be cheaper than any other type of construction machine—with one exception. That exception is when you have to move the machine frequently. Maximum ground speed for an excavator is just a little over 3mph, while dozers might be 7mph and a wheel loader might be 25mph. But an excavator only burns maybe 6gph while a dozer or a wheel loader might burn *double* that. So if the machine needs to be moved frequently, there is probably a more efficient machine than the excavator—but that's the only handicap for this machine."

LEFT
Richard Venne, a Pacific Blasting excavator operator, uses the Hitachi 270 to dig into the still-warm carcass of the Pacific Palisades hotel structure within an hour of the series of explosions that dropped the structure. The 270's cab has been retrofitted with guards to protect against falling objects. The cab itself, like those on all machines sold in the United States and Canada, provides roll-over-protection (ROP) for the operator through extremely heavy structural components at each corner. The cab is strong enough to support the inverted weight of the machine—an extremely rare requirement.

One out of three construction machines sold in the United States today is an excavator; ten years ago it was one out of ten. "That's because, first and foremost, excavators have low owning and operating costs," Murphy says. "Anything you can do with an excavator will be done cheaper than with any other type of machine—except where mobility is a factor."

Although an excavator's mobility around a job site is low, it works very fast at a multitude of demolition jobs. Within its

25

Cat's Model 325 is a mid-sized machine weighing about 61,000lbs and powered by a 168hp computer-controlled diesel engine. Top travel speed is about 3mph.

reach, an operator can scoop, push, pull, prod, grab, or hold tools. With a hydraulic breaker attachment installed in place of the bucket, the operator can reach into otherwise inaccessible places to smash concrete from the underside of a damaged freeway overpass. With a "grapple" type bucket the operator can grab steel I-beams from a pile of rubble and move them to a pile of scrap metal for recycling. A common bucket's teeth provide a set of strong, finger-like extensions that an operator can use to precisely manipulate one component of a structure—pulling on the corner of a brick wall, for example, in just the right place to bring the bricks down in a controlled pile. A skilled operator, like Banderas Wrecking's David Thompson, can use the bucket and the weight of the machine to neatly fold a steel stairway into a tidy, compact bundle of scrap steel ready for recycling.

The key to this economy and precision is in the hydraulics and in computer control. "You can set the machine up to do the job you need to do, without wasting fuel through excess engine- or hydraulic-capacity," Mike Murphy says.

"The demolition industry is a tiny little blip on the marketing screen for excavator manufacturers today," Murphy reports, "but it clearly is going to be much bigger in the future; growth of this industry is dramatic. Of the approximately 12,000 excavators sold in the United States during 1994, only 200 to 500 went directly to the demolition industry.

Excavators on Demolition Jobs

Watch an excavator working on a construction job, then watch the same model work a demolition project: the difference is amazing. An excavator busy digging trenches for drainage tiles, for footings and foundations, or any normal construction project leads a predictable and boring life—dig, swing, dump, return. The cycle repeats all day, about three times a minute. The most excitement for the operator comes when you get to move the

You need a key for the Cat 325, just as with a car. Above the key is the power selector, here set for level five.

upper structure of the excavator to the right or left. With the boom up at about a 45deg angle (right hand, pull back—remember?) and with a couple of feet of ground clearance for the bucket, use the left joystick to move the stick forward and aft. It is really simple.

Now, adjust the stick so it is straight up and down, with the bucket a couple of feet off the deck; you might need to adjust the boom a little. This is the travel position. Together, the boom, stick, and bucket function a lot like your upper arm, forearm, and hand, and as with your arm and hand, you can reach, grab, pull, and push objects. The real difference is that your reach now extends to about fifty feet, and you can apply up to 25,000lb of force to objects.

So far, your moves have been simple and straightforward, but that is about to change. With the bucket still up and in close to the cab, move the left joystick gradually to the left; the upper structure will begin to swivel to the left. The control is progressive, smooth, and predictable. Ease off on the stick and you can line up pretty much exactly where you want. Now take a look around the machine to make sure nobody has snuck up on you and, when all-clear, move the joystick to full left

deflection. The upper structure pivots quickly to the left making a full revolution in about five seconds. That is tremendous agility in a machine weighing over 60,000lbs.

Directional Controls

Finally we will demonstrate the pedal/lever combination. These move the machine around the job site. Each control a track; with the bucket in the travel position, up off the ground and with the stick hanging straight down, push the left pedal slowly forward and the left track will begin to move forward. Press harder and the track speed will gradually increase. You now have the machine rotating slowly around to the right. Let back on the pedal and the track speed slows, then stops as you reach the central position for the control. Now take your foot off the pedal and try the same trick with the lever—it works exactly the same way. These two controls feed hydraulic pressure to track motors on the undercarriage. They are completely independent of each other, permitting tremendous maneuverability. Now push both levers forward together, gradually; the motors smoothly apply power to the tracks, and now you're moving forward at the blistering speed of about one-mile-per-hour.

Come to a stop, then pull back on the levers—now you're going backwards. The control is intuitive, smooth, seamless, and *slow*. Maximum speed forward or reverse is about 3mph. If you can drive a kiddy car you can drive one of these things around. Okay, since you think you're so hot, split the levers, one forward and one back, full deflection; now the 325 pivots on its own axis. It isn't a trick or a stunt—maneuvering on a job site requires even thirty-ton machines to turn on a dime to get into position for a "hook" or to work around obstacles.

Getting to Work

While each of these controls is simplicity itself, putting them all together in a seamless ballet of mechanical grace is a little more complicated. It is about as complicated as driving a car and takes about as long to become reasonably proficient.

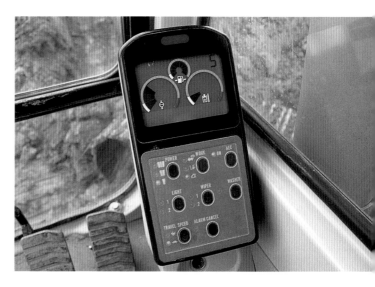

RIGHT
Conventional gauges have been replaced by a computer-driven display—rather like the "glass cockpit" of the F-15 fighter, without the indicators for guns and bombs. Instead, you get information about fuel quantity, hydraulic pressure, engine power, and other controls. "AEC" is automatic engine control, a system that, when engaged, will return the engine to IDLE after four seconds of control inactivity.

BELOW
Jim Hasty pivots the 325 on its vertical axis by splitting the tracks, the left one at full FORWARD setting, the right track full BACK. It's a handy way to maneuver in tight spots, a kind of agility not expected from such a big machine.

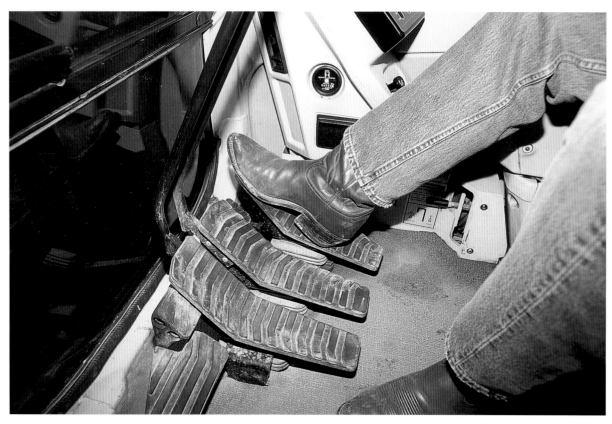

Jim can control a bucket clamshell or hydraulic attachment with the far-right pedal control.

Well, let's waddle over to the rubble pile and see how it works in the real world. Sticks forward and off we go. There's part of a brick wall up ahead; somebody's already taken it half way down, and it isn't providing support for any part of the building nearby so let's try hooking a section. Maximum reach with our boom is 38ft, but at that distance you can't apply much force with the bucket, so let's maneuver closer, to about twenty feet, where we can get some leverage. We have to squirm and wiggle over and around the rubble to get into position—now you can appreciate the tremendous agility the directional controls and the independent tracks provide.

Okay, we're out of range if the whole wall decides to topple this way, and we've checked to ensure nobody's working nearby. So let's reach out with the bucket and take a bite out of the wall: *boom up, stick out, bucket dump* with the joysticks, all together in a single, smooth move. Hmmm, well, try one at a time—you'll get it, sooner or later. The bucket will swing upside down, teeth out like big fangs. Maneuver these teeth into position at the top of the wall, about five feet from the end. You can, if you are careful, put the outboard tooth exactly on the center of the wall, between two bricks. Lower the boom and adjust the stick together, applying just a little pressure to the wall to get a grip on the structure. Try moving the bucket in and out, to the left and right, just a bit. You can see the brick wall flex, then cracks appear in the mortar as a section weakens. Reposition the tooth to the outside of the wall, then just pull the section out of the wall; the bricks peel away, fall to the ground with a cloud of dust.

36

David Thompson at the controls of the Komatsu PC220 LC-6. Although air conditioning is optional on all modern excavators; most operators just open up all the windows and door. The windshield slides up and locks overhead, out of the way. That slender lever with the red ball on top is the control lock, seen here in the unlocked position. The engine will not start unless the lock is first placed in the aft position, ensuring no dramatic surprises when the hydraulic pressure starts coming up after you light off the powerplant.

Generous track shield keep the bigger chunks of concrete out of the track rollers and the other tender bits of the undercarriage, prolonging component life and reducing down time.

Now, boom up, stick up, and move forward toward the remaining portion of the roof, hanging limp above the second story. You can do all three maneuvers by using the pedals for maneuvering over the rubble while you elevate the bucket into position for a hook on the roof. Sink your teeth into some of that tar paper and gravel just on the far side of a beam (bucket close, boom down, stick in) until you've got good purchase on the material, then bring the stick in by easing back on the joystick. The structure will tremble and the walls will wobble a bit as a big section of the roof comes out. The wooden rafter splinters and falls to the ground in front of you, followed by fragments of roofing material and a few bricks from the ends of the rafter. That's excavator work on a demo job.

Excavator Attachments

The demolition industry has been transformed by many large and little technological innovations over the past decade or two; the development of hydraulics was one of these, the addition of computer control was another, and a third was the development of stick-end attachments for excavators. These attachments have revolutionized the way buildings are dropped; with the speed and precision of hydraulics, the control of computers, and the amazing power of demolition attachments, contractors execute jobs with speed, economy, and safety impossible twenty or thirty years ago.

Attachments use the excavator's hydraulic pump to energize systems of extreme simplicity and stunning power—shears that can slice through a 12in steel I-beam, hydraulic breakers that can shatter huge concrete piers, and grapples that can grasp structural elements in a 35,000lb grip.

The development of these attachments came right out of the demolition industry, from a little father-and-son company called Allied Wrecking & Dismantling. Back during the 1970s, the company won a bid on a project that included a very large quantity of copper bus-bar material. These bars needed to be cut up for transport. They started out with a small commercial shear; it worked, but not quite as well as they thought it could, so a home-brewed machine was cobbled together and bolted to a 580 Case backhoe-loader. This improvised shear wasn't a lot more than two pieces of thick sheet steel with a simple

Those big hydraulic lines contain about 5,000psi, enough power to provide the operator with around 30,000lb of bucket curl force.

RIGHT
Here's one reason demolition work is considered severe service for heavy machines—a mixture of dirt and concrete that is far more abrasive than the kinds of soil typically encountered by excavators on construction projects. The full-length track guards help keep the biggest pieces out of the works.

pivot and a hydraulic cylinder providing the power. The steel certainly wasn't tool grade material, but the home-built attachment allowed the little Case 580 to cut four bus-bars at once rather than only one at a time.

They decided to try scaling the design up to fit on the stick of a model 6400 Link Belt excavator; this new version was intended to cut steel. As one of Allied's senior project managers, Jack Stewart, explains, "When you're tearing down buildings without shears, you've got to use torch men for everything. It's expensive, finding people who are good with a torch is hard, and there are all sorts of other problems with the old method." The new design was fabricated by a machine shop in Allied's home-

David Thompson uses the Komatsu 220 to fold a steel stairway up into a tidy ball of scrap metal small enough to fit in a haul truck. With a power-to-weight ratio approaching one-to-one, the excavator easily applies much of its weight to the mass of scrap metal, pressing it tightly together.

town, Youngstown, Ohio; this time it was complete with tool-grade steel blades.

"I was chief engineer for a fuel company in the area at the time," Jack recalls, "and they were doing some demolition for us. It was fascinating to watch them—nobody had ever seen anything like it. They could just reach up with that excavator and cut down pipelines with no problem at all."

Allied Gator was formed in 1981 to market the shear. The originals were fabricated from sheet steel, which tended to crack at the welds. The design was modified, and the shears are now assembled from large castings of an extremely tough steel, the same basic material used as part of the armor on M1 "Abrams" main battle tanks. The new design has completely eliminated the cracks and breaks of earlier designs. Allied makes three sizes of shears for excavators from 50,000lb and up. The biggest will slice through a 48in beam in one bite!

"Everything we've built has been, first and foremost, for our own demolition work," Jack says. "We try our products on our jobs first, and we design the kind of tools that we want to have on our jobs—products that last a long time, don't need much maintenance, and will cut through any material we're likely to find on a project."

The ideal excavator for use with a mammoth shear like Allied's is a high-pressure machine like Hitachi's EX 1100, the Komatsu PC 1000, or Caterpillar's biggest, the 375, all weighing in at nearly 200,000lb. All these machines provide more than 4,000psi hydraulic pressure. The normal method for mounting the shear is on the boom of the excavator, not on the stick. Even with the big excavators, the weight of the shear restricts maneuverability. And even with the biggest excavators, some modification of the hydraulics is part of the installation routine because it takes the full output of both pumps to

Normal position for the track drive motors is at the rear of the machine; when the operator gets things turned around like this the travel controls are reversed. Normal position for the tracks is on the ground, but this maneuver is fairly common, too.

operate the big shears with reasonable speed—and even then, a shear operating cycle can take up to fifteen seconds or longer.

Claw Bucket

The claw bucket is another Allied innovation, developed as a result of another common problem on demolition jobs, the clutter of steel and other bulky material. These beams are relatively easy to break out of the structure—but then what do you do with them? They get in the way of the machines, slowing down the whole process. The claw bucket, or grapple, was a way to deal with this unwieldy material.

"The claw has two advantages over any other bucket cover," Jack Stewart says. "There have

been demolition bucket covers around for years, but most don't have the strength to stay closed if you try to invert the bucket. Ours are made with so much power that we can pick up twenty-five tons of scrap and turn it upside down to shake the dirt out without spilling any of the scrap. That's important with steel scrap because the recycle plant penalizes you for the amount of dirt in a load; with clean scrap there isn't any deduction."

Hydrocrackers

Hydrocrackers, or "breakers" as they are often called, use the hydraulic pump of a machine to operate a kind of huge jackhammer. Some of these tools are just immense; Teledyne's TB-1624X weighs nearly 6,000lb and delivers up to 450 powerful impacts per minute.

With one of these things out on the end of your boom or stick, you can smash a precast concrete wall from a tilt-up building into bite-sized chunks in no time.

Breakers are really impressive to watch in action. The biggest—like INDECO's Model MES 12000—weigh more than a small excavator and deliver up to 16,000ft/lbs of energy with every cycle. That is an incredible amount of power. These machines are huge, with price tags to match, but they typically only have two moving parts! They are extremely useful, popular tools that have been a big part of the revolution in the building demolition industry's bag of tricks.

Concrete Crushers

Among the worst jobs in demolition has traditionally been the problem of breaking up concrete. Alone, the stuff is hard and tough; you can break it up with a crane and ball, but it's a slow,

dirty, job. And once concrete is reinforced with rebar, it is particularly hard to work with. But a properly equipped excavator can crush rather than cut material.

TRAMAC offers three models from about 2,200lb to 9,200lb for stick-mounting. These crushers can all rotate, like a hand with 360deg mobility.

Crushers are manufactured in several configurations, some able to cut as well as crush, others strictly crushers. They are excellent for jobs like earthquake-damaged freeways and many have been used to clean up the mess left by the Loma Prieta and San Fernando quakes in California.

LaBounty Manufacturing sells a version of this design they call a "Universal Processor." These attachments work with small excavators of about 5,000lb and skid-steer machines in the 4,300lb class; interchangeable jaws allow fast processing of concrete and metal for interior and exterior demolition.

Caterpillar Excavator Line (1995)

Model	Horsepower		Operating Weight		Max Reach		Max Depth	
	(kw)	(hp)	(kg)	(lb)	(m)	(ft)	(m)	(ft)
E70B	40	54	6,900	15,200	6.7	22	4.6	15
311	59	79	11,390	25,100	8.1	27	5.6	18
312	63	84	12,340	27,200	8.6	28	6.11	10
320	95	128	19,120	42,150	10.6	35	7.6	25
320 L	95	128	21,300	46,960	10.6	35	7.6	25
E240C	110	148	23,000	50,705	10.6	35	7.4	24
EL240C	110	148	23,600	52,028	10.6	35	7.4	24
325	125	168	27,670	61,000	11.5	38	8.2	27
325L	125	168	28,120	62,000	11.5	38	8.2	27
330	166	222	32,660	72,000	12.4	41	8.9	29
330L	166	222	32,700	75,000	12.4	41	8.9	29
350	213	286	50,800	112,000	13.5	44	9.6	32
350L	213	286	51,710	114,000	13.5	44	9.6	32
375	319	428	81,650	180,000	16	53	10.9	36
375L	319	428	84,820	187,000	16	53	10.9	36

The 992 is the John Deere Company's flagship excavator, a huge, million-dollar machine with a long reach and a powerful bite. The E-model weighs nearly 100,000lb and can dig a hole 30ft deep or pull down a roof truss 38ft from the center-post, although bucket curl force out there is reduced from the stunning 52,000lb possible with the machine. The 992 is propelled by a 285hp fuel-injected, computer-controlled engine—and does the standing quarter-mile in just five minutes.

45

Operator Alex MacLeod collects one of optional buckets for the machine and heads for safer territory while the crew from Pacific Blasting and Dykon are busy stuffing dynamite into the bore-holes in the Pacific Palisades hotel. The 992, like the Hitachi 270, has heavy armor for the operator.

RIGHT
A lot of demolition work involves moving material around once the building has been dropped. These bricks were recently part of the wall that landed on a section of roof. The excavator operator deftly scoops the bricks off before hooking the roof support.

46

David's assignment on this job is to sort out the rebar and scrap metal from a seventy-year-old grain elevator being dropped by a ball and crane. While Ernie Asher pounds away with the "headache" ball nearby, David takes the Komatsu up the rubble pile to pluck out as much rebar as he can reach. The 220 is roughly equivalent to the Cat 325 in basic specifications and performance.

This freeway was built about 1950 and damaged beyond repair in the 1989 Loma Prieta earthquake. Here's an excavator breaking up the underside of the structure with a sharp, heavy steel spear-type tool, just one of the many kinds of attachments available for demolition work with excavators.

RIGHT
Here's another excavator attachment, a concrete crusher. This device will nibble away at the concrete, separating the rebar.

BELOW
Take a close look at this because you don't see an excavator this clean very often. This one is a Caterpillar 245 just delivered to a demolition site. If it looks big, it is—about 145,000lb of metallic muscle. Maximum reach for the machine is a whopping 46ft and that hefty boom and stick can lift 33,000lb.

Excavators perform a wide variety of tasks on the job site. Here, a large stainless steel tank is being loaded into a debris haul truck. The tank, along with all the metal from this job, will be recycled.

RIGHT
Taking rebar out of the rubble is a little like weeding a garden—but these weeds are tougher and more tenacious than anything in your lawn. The excavator is about the only machine that can efficiently pull this material out of the debris.

The front-end loader—tracked or wheeled—is an essential member of the wrecking team. A tracked front-end loader looks a lot like a bulldozer with a big, mobile bucket instead of a blade. While the excavator nibbles away at the building, the front-end loader rumbles around behind cleaning up the mess.

Most loaders you'll see on demo projects will be tracked. The wheeled variant is faster, but the "road hazards" on a demolition site eat tires almost as fast as you can replace them. But there is a place for them on some jobs, particularly large industrial sites, structures without basements. They are much faster than the tracked version, with some models able to reach 30mph instead of the 6mph top speed typical of the tracked loader. That speed really makes a difference when you're cleaning up a 400-acre industrial site, where large volumes of material need to be moved longer distances.

LEFT

The Caterpillar 977 series track loader goes back to 1955. The L Model shown was built from 1978 to 1982, and this example seems to be in pretty good shape considering its age and the kind of work that track loaders perform in the demolition industry. The 977 is helping with the final stages of a project in Raytown, Missouri— dissembling a rather flimsy single-story shopping center that was about to fall down all by itself.

This clean-up on a demo job is a constant process. The front-end loader operator continuously separates material for recycling—steel rebar, pipe, I-beams, copper wire, electrical motors, aluminum sheet—and the concrete, brick, and wood rubble destined for landfill. The rubble gets crushed under the loader's tracks, scooped up, and delivered to the pickup point where another front-end loader will probably be busy loading the debris trucks that form a constant parade through the job site.

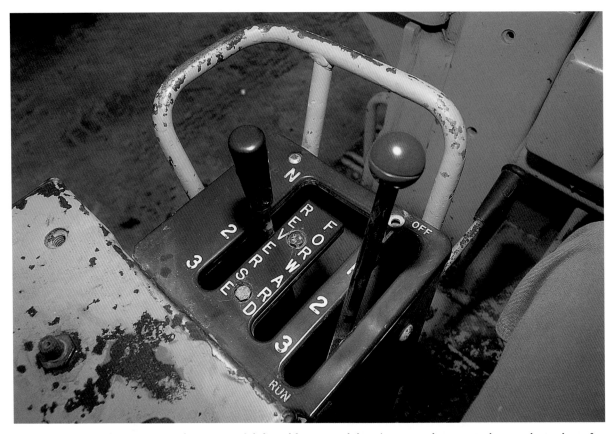

All the track controls are over on the operator's left and front—and there's not much to control: one selector lever for three speeds forward or reverse, and a throttle lever with three speeds and an OFF position.

While a track loader might look like a bulldozer from a distance, it sure has a different job to do. A dozer pushes material around, the track loader scoops it up and pours it onto a pile or into a truck. So instead of a semifixed blade, the loader has a mobile bucket with a wide range of movement.

A typical track loader like the Cat 935B (and Cat dominates the loader market in the United States) tips the scales at a bit over 33,000lb and can manage a bucket capacity from 2 to about 2.4 cubic yards.

Tracked loaders have been around for quite a while—not as long as the basic dirt-pusher, but Cat was selling the HT4 as early as 1950. That old relic was a 54hp machine that

The left pedal controls the left track clutch and brake. Jim Hasty from Holt Brothers is at the controls.

lasted into the 1980s and grew to 35,000lb wet with a two-yard shovel.

How To Run A Track Loader

While new and old excavators are pretty much the same, track loaders changed significantly. Older models, like the 955s found in abundance on demo sites around the country, are simple, straightforward, mechanical systems that require the operator to manipulate the controls to do just about everything.

The new machines, though, automate much of the process, easing the workload of the operator and increasing productivity a bit. The new designs, too, move the engine to the back of the machine, providing a counterweight to the bucket load and making the machine more stable with a load.

So how do demo contractors and operators like the new designs? Well, a lot prefer the old models. They are less likely to break, are simpler to repair, and far less expensive to purchase and operate.

Inspection

Just as with any other piece of expensive machinery, the operator is supposed to make a careful survey of the equipment before climbing aboard and starting the engine. As with the excavator, slowly walk around the machine, checking for loose, damaged, or missing components; check the fluid levels—you should have full fuel, oil, and coolant. Hydraulic fluid should be checked with the engine running and the system warmed up.

Find the master electrical switch and turn it to ON. Then up into the seat, put the key in the ignition, and turn to ON. Start sequences for newer and older track loaders vary. The older models still found on many jobs require you to select the pre-heat position for the engine for a few seconds to heat the combustion chambers, then turn the control to the far right START position; the engine will crank and should fire up. Release the switch and it returns to the central RUN position. Let the engine warm up at idle for about five minutes. When it is quite cold, the engine needs to warm longer.

Push forward on the inboard lever to lower the bucket and arm assembly.

weighed in at only 5,500lb wet with a skimpy little 1.25 cubic yard bucket. But the idea was obviously a sound one as bigger, stronger loaders followed almost immediately: the 955 series

The production cycle begins with the LOAD part of the sequence. Here, an operator from Spiritas Wrecking and Demolition digs into the rubble left behind the Case 980 excavator. He will approach the edge of the pile square-on, bucket tilted very slightly forward, and will load by driving forward and rotating the bucket to scoop up about 3 cubic yards of material.

The Loader Work Cycle

A loader's work routine is based on a cycle of events that repeat over and over, all day long.

Loading

The routine begins with loading. Normally, that involves driving the loader up to the pile of material, approaching so the bucket is square to the face of the pile, just clear of the ground, with the bottom of the bucket slightly tilted forward. The operator drives the machine forward, rotating the bucket at the same time to scoop material from the pile—aiming to fill anywhere from 60 to 100 percent of the bucket, depending on the type of rubble. Actually, you can pretty well forget about filling 100 percent of the bucket on demo jobs because most of the rubble is in the form of rather large chunks of fractured brick, concrete, or cinder blocks. But that's the *load* part of the cycle and it should take from two to twelve seconds, depending on the material.

Maneuvering

Maneuvering with the load typically involves four changes of direction, two at each end of travel. This part of the sequence will routinely take about another twelve seconds.

Travel

Next, consider the travel time of the machine from the rubble pile to the "target," a truck or a debris pile where the material will be unloaded. Obviously this can vary tremendously, although the trucks will usually be brought as close to the work as feasible. On demo jobs, travel time will often account for fifteen additional seconds for the production cycle.

59

That big chunk of concrete would be an awkward load for an excavator's standard bucket, but it's a piece of cake for the front loader, shown here with the bucket raised to the carry position. With the bucket up like this, the operator has excellent visibility and the load is ready to dump into the "target."

Dump

The operator will raise the bucket to the carry position while maneuvering away from the debris pile and while traveling and be ready to dump just as soon as the "target" is in range. If the target is just a pile of segregated material, dump time approaches zero; for a smaller target like a truck, the operator will be more careful and this portion of the cycle will take two seconds or so.

Loader Productivity

Add it all up and a typical demolition project loader will average about one cycle every thirty-five seconds or so, or around 103 cycles per hour. If you're using a 2-yard bucket that averages, say, 75 percent full, you can expect to move about 155 yards of material per loader per hour, or 1,236 yards per eight-hour shift. Demolition contractors need to be able to

Bobcats are a smaller, more maneuverable type of front-end loader.

accurately gauge how quickly their equipment can work, at what capacity, and do all the math required to figure out a bid for a job that is competitive yet allows the contractor to make money. Math problems are a big part of the demolition business!

Loader Economics

Caterpillar's wizards have looked into the corporate crystal ball and guess-timated what it actually costs to buy, own, and use a machine over a unit's normal lifetime, a figure for loaders that varies from only $12 per hour for the smallest machine, the 933, in moderate service, up to $68 per hour of operation for the biggest machine, the 973, in severe service. That includes purchase, fuel, repairs, maintenance labor and parts but excludes operator wages, insurance, and interest on the bank loan.

Loader buckets experience extremely severe service, resulting in high rates of wear on the teeth, slides, and bottom surfaces. Replaceable wear-plates and teeth at these critical points take most of the abrasion and permit much longer bucket life.

63

This 977 has cleaned up the site right down to the dirt. Here, the operator is separating and preparing debris for loading on the haul trucks.

Here's a good look at the ROP (roll over protection) cage that is now mandatory on all loaders and construction equipment sold in the United States. The four steel beams surrounding the operator will support the weight of the machine—about 48,000lb in this case—if it should tip over on a slope.

The little "backhoe/loader" is a lightweight machine with a heavy punch. This 580K from the Case company has a hydraulic breaker installed on its little stick and is hard at work breaking up some of the nastier bits of the dear departed Pacific Palisades hotel. Notice that the operator has used the machine's bucket and outrigger to position itself in space. That big John Deere 992 in the background could do this job, but the 580's daily rental rate is a tiny fraction of the bigger machine's.

68

A large grapple on a Cat 973. The grapple bucket makes short work of wood frame construction homes and businesses. The grapple's gaping maw is designed to grab two to three cubic yards of splintered wood, wall-board, plywood, and electrical conduit.

Although front-end loaders are huge, dramatic, and costly, the basic skills involved in driving them are surprisingly simple. Operating them skillfully is another matter; it takes years of experience to efficiently use one at a demolition site.

A little 753 Bobcat is just the ticket for cleaning up recyclable materials. It is small enough to work inside structures. It is fast, agile, and provides the operator with a little armored turret to hide in when the roof comes down.

PREVIOUS PAGES
An experienced crane operator can hit a mark on a building with great precision. That's important because this kind of demolition work requires control for safety.

RIGHT
Ernie Asher operates the crane's controls, which control the position of the boom, the cable, and the upper body of the crane through a system of mechanical clutches and linkages.

Concrete gets harder as it ages and this concrete has had seventy long years to toughen up. Even so, this "headache ball" rig is more than a match for this old grain elevator complex in North Kansas City, Missouri.

RIGHT BELOW
Without those outriggers, the crane would be pretty easy to tip over. Even with the outriggers, it still can be done and the operator has to maintain "situational awareness" at all times to keep from doing something embarrassing.

The ball weighs about 5,000lbs. Ernie swings it by pivoting the crane to the right, then at an artfully chosen instant, back to the left. The impact of the ball smashes through six inches of reinforced concrete and large chunks thunder to the bottom of the inside of the structure.

5,000lb of cast iron meets six inches of cast concrete; the iron wins every time. Crane-and-ball on this kind of job is a slow process, but a precise one—and the only practical way to drop such a structure.

five

Explosives

Early Sunday mornings in Vancouver, British Columbia, are normally quiet and peaceful. This particular November morning, though, was full of an expectant, electric tension—even at 6a.m. as I drove from my hotel to the Pacific Palisades hotel. A gentle Canadian mist fell on nearly abandoned streets—the only things moving were tow trucks carting away the cars of people who apparently couldn't read "no parking" signs. The only indication of what was to come were the battalions of police awaiting the orders to shut down the streets, the amber lights on their cars flashing a warning in the predawn gloom.

I parked the car and walked to the hotel adjacent to the job site. The lobby was full of people, all apparently overdosed on adrenaline. I took the elevator to the twenty-second floor, and set up in one of the rooms rented by Pacific Blasting, the prime contractor for the job. Through the hotel room window, I could see crowds gathering on the streets, roof tops, offices, and hotel rooms surrounding the doomed building.

The Pacific Palisades Hotel was only about twenty years old, a reinforced concrete building built in the mid-1970s. It appeared modern, sound, and perfectly functional except for one flaw—it was in the way of a bigger, more profitable structure planned by a developer. So, at the tender age of twenty, the

LEFT

The Pacific Palisades hotel, Saturday morning, November 2, 1994. The doomed structure, only about twenty years old, is loaded with about 200lbs of dynamite. Despite the urban setting and the proximity of other buildings, some only 20ft away, the structure will be dropped with explosives.

A fresh box of high-quality Canadian dynamite. The material comes in several shapes and strengths. This is Power Primer, a seventy percent formulation in 1.25in-by-8in sticks that is manufactured by Continental Explosives.

Pacific Palisades Hotel was about to die. It had taken a year to design and another year to build, but at 8 a.m. someone down below would push a button and the building would collapse. Considering how long it had taken for the building to be born, it would die with remarkable speed.

Three companies worked on the blast, Pacific Blasting, Demolition & Shoring (prime contractor), Continental Explosives (a Canadian explosives vendor), and Dykon (an implosion demolition specialty firm from Tulsa, Oklahoma). Pacific provided the "heavy lifting," managed all the permits and subcontractors, cleared the building down to the con-

crete. A few days before the scheduled shot ("shot" is industry slang for a structure demolished with explosives), Jim Redyke (founder and chief guru of Dykon) traveled to Vancouver to "design" the shot.

Planning

"Bringing down a building basically involves removing vertical supports—the columns—in a controlled, sequential way that then uses gravity to collapse the structure," Jim Redyke says. "You can control the direction of fall by taking out supports that force the building to fall where you want it to go…kind of like taking one leg away from a three-legged stool. The weight of the structure and gravity does the rest."

While the theory is simple, planning a successful drop is quite complicated. The shape and composition of the columns has to be studied and tested. Core samples are taken, the original construction drawings studied, and at least one test "shot" will be fired to verify the calculations.

Although less than 200lbs of dynamite and detonating ("det") cord will be required for the Pacific Palisades building, those charges have to be placed with great precision to be effective. The basic idea is to weaken the columns on one side of the building's lower floors, starting at the bottom and working upward over a period of about ten seconds. Each charge will cut through the concrete of a column and the weight of the structure above will start the collapse.

Part of the art of implosion demolition involves slowing the event down into many small, calculated blasts instead of one huge explosion. That is accomplished with time-delay blasting caps that will initiate the dynamite and det cord over a period of several seconds.

The explosives will shatter the concrete around the reinforcing rods—and since the concrete provides nearly all the strength of the column, that part of the building quickly begins to fall. If enough columns are shattered the building will collapse. This much is easy. The art of demolition is knowing which part of the building to take out at each moment over a period of ten seconds or so; errors in this kind of calculation can be rather embarrassing, especially if the structure comes down where it isn't supposed to.

Explosives

Explosives come in all sorts of forms—gels, granules, powders, cord, liquids, plastics (in blocks and sheets), and old reliable, stick dynamite. All have properties designed for specific conditions. Huge quantities are used every year, often in urban areas and often without anybody even noticing the detonations.

Dynamite is a mixture of nitroglycerin, a liquid, and a binder. It was the first practical high explosive and revolutionized mining and construction by making the blasting process safer and more efficient. As everybody knows, dynamite is sold in sticks, typically 1.25in by 8in. As few people know, dynamite is rather insensitive and difficult to initiate. You could probably whack it with a hammer and survive, but don't try it.

Dynamite won't detonate unless "initiated" with a priming charge, normally from a blasting cap. While some blasting is still done with time fuse and suitable nonelectric caps, virtually all construction and building demolition blasting today uses only electrical caps, fired by wire from a remote location. That means that you can wire 100 charges into a big firing circuit and fire them all at the same instant with a single push of a button. Timed detonating caps allow you to press that same single button and stagger your single explosions by ten seconds or more.

Rather than firing all the charges at once, Jim Redyke designs a shot to evolve over a period of ten or fifteen seconds. That's possible because blasting caps are now available with built-in and extremely accurate tiny fuses that permit delays of ten or more seconds. For a tall building like Pacific Palisades, Jim will have the charges on the lowest floors and in the basement fire first, chopping the base from under part of the building and leaving part intact to act as a kind of hinge. The weight of the structure will begin to pull the building down in a controlled direction. The remaining charges fire at preset intervals of about one second, frac-

Stan Holtby primes a stick of dynamite with an electrical cap. This one happens to be an instantaneous type, but others will use tiny fuses to delay the explosion for up to several seconds.

turing the structure's internal supports, weakening it from the inside out. Then, as it falls, the once strong structure's own weight tears it apart leaving nothing but a pile of pulverized concrete and reinforcing rod.

Test Shot

Early in the design sequence, Jim will find a column, normally in the basement, for a test shot. Based on a core sample and available information about the nature of the material in the column, he'll mark locations for placement of explosives. The crew with the drill bore the holes to Jim's specification, usually dead center and almost all the way

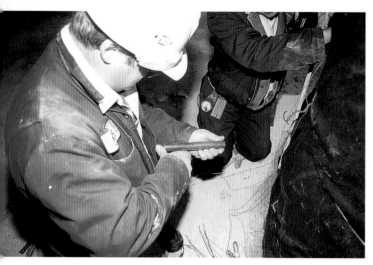

Corry Goumans cuts open a stick of dynamite with his pocket knife. The slit will allow the stick to deform and expand when the charge is tamped into the bore-hole and will help maximize the efficiency of the charge.

through. On a job like Pacific Palisades, Jim might tell the crew to bore four holes in the test column, each deep enough for four 8in sticks of dynamite. Then, after receiving authority from the city for the shot, the holes are loaded and fired. Deep inside the building, the blast's noise and "fly rock" are fully enclosed; people nearby probably don't even hear the detonation. Then the Dykon crew re-enters the structure to inspect the damage. The column should be completely shattered, although the rebar will still be intact; if the column isn't demolished, more holes and more explosives are required.

So, how much dynamite does it take to drop a twenty-two-story building? Not much, if it is placed correctly. Jim's design for the Pacific Palisades building uses a bit less than 200lbs, plus a small amount of 'det' cord. Actually, gauging the amount of explosive to be used is key to a successful drop. Dynamite is cheap, about $1 per stick, so the cost of the material isn't a factor. The trick is using enough to be sure that the building comes down exactly where it is supposed to without excess flying debris ("flyrock" in the blasting trade) or breaking windows with the noise of the shot.

There is a real art to the business and some do it better than others. If the dust clears from a shot and the building is still standing, it's more than embarrassing—the standing building is now a disaster waiting to happen, weakened by the blast and threatening to fall at any time.

Once the test shot confirms or refines Jim's understanding of the structure, he walks through the building with a spray can, marking the location for each bore hole. Then the drillers come along, bore the holes, then mark each with a length of red-painted rebar. Other people from the contractors' team will wrap exposed columns with a special fabric used in construction, then enclose all with chain link fencing, leaving the red rebar exposed to mark the bore holes.

Finally, the day before the shot, loading begins. Corry Goumans, blasting superintendent for Pacific Blasting (prime contractor) and Stan Holtby (regional manager for Continental Explosives), meet Jim Redyke on the site on Saturday morning. A guard is posted on the site, close to the two big red "powder" magazines, one containing the electrical blasting caps, the other with the dynamite and det cord.

Stan, Corry, and Jim inventory the caps and dynamite carefully, then begin loading. All handle dynamite on a nearly daily basis and they work quickly, methodically. The rebar is pulled from each hole, a cap with the correct delay element selected, then the first stick in the hole is primed and inserted in the hole.

Priming and Loading

For most people, whose notion of dynamite is that it is extremely sensitive and dangerous, the priming process is full of surprises. It begins, for example, by taking a stick of dynamite and poking a hole in it with a sharp tool. Into the hole goes the blasting cap, a bright aluminum tube with a long pair of wires attached. Stan extends the wires, ties a quick pair of half-hitches around the stick, then inserts it in the bore hole. Then, Corry uses a wooden tamping rod to push the stick into position at the bottom of the hole. Two other sticks follow the first, each slit with a pocket knife before

"Jeeze, Stan, do ya' think this is a good idea?" Corry Goumans, blasting superintendent from project prime contractor Pacific Blasting, watches as Stan Holtby punches a cap well in a stick of dynamite with one leg of the cap-crimper tool. The column has been wrapped with chain link fencing material and Tyvek® construction fabric to minimize "flyrock." Red paint marks the bore-holes to be loaded with explosive.

insertion. Corry presses each into position with enough pressure to expand the sticks (or "cartridges" as they are sometimes also known)—rupturing them, in effect—so they form a good contact with the walls of the bore hole. The exposed ends of the wires are twisted together as a safety measure until the firing circuit is connected later. But first, to finish off the bore hole, Stan inserts the nozzle of an applicator in the hole and releases a bit of foam. The material quickly expands, then hard-

ens, forming a seal that will concentrate the force of the explosion. Without it, the charge would squirt some of its energy out the hole, drastically reducing the effect of the explosives.

Once all the charges are loaded, the wires are spliced together into a big electrical circuit. Stan, Corry, and Jim test for continuity, and then they are done, except for one difficult part…the waiting. Until the building is dropped, nobody will get much sleep.

Fire in the Hole

At precisely 7 a.m. on Sunday, November 3, 1994, the patrol division of the Vancouver Police Department shut down every intersection within a half mile of the Pacific Palisades building. At the same time the doors of all adjacent and nearby buildings are locked. The streets are utterly deserted. For the next hour several thousand people fidget nervously, joke, and jockey for viewing position on all the rooftops and windows around the structure.

The early morning drizzle stops, the fog clears, and the sky begins to brighten. Two hundred yards from the building, out in the middle of the street, a VIP and media tent shelters a few favored people. A feature film crew put the finishing touches on two patrol cars parked a few feet from where the rubble is supposed to land; the collapse of the building will be worked into a story. With the cars positioned, red lights flashing, the crew retreats to their camera position.

There are, in fact, about a thousand camera positions around the Pacific Palisades including motion picture cameras in armored housings very near the building. They will be operated by remote control, recording the demise of the hotel from only about 100ft away. But the crowd of spectators is heavily armed with photo artillery—video cameras, Instamatics, Super 8mm and 16mm film cameras, and 35mm cameras in every brand and configuration.

7:55 a.m. Quite suddenly, and in perfect synchronization, the sirens of all the police and fire vehicles assembled around the site wail for about twenty seconds, then, just as suddenly, stop together. The silence is deafening as the echoes of the sirens die away.

Stan and Corry methodically load each bore hole. The wires for the electrical blasting caps will be connected in a big circuit, carefully tested, then fired all at once. The charges will detonate progressively over a period of about ten seconds.

RIGHT
Standard procedure for Dykon is to conduct a test "shot" of a column before determining exactly how much explosive to use, and here's the result of the test in the basement of the Pacific Palisades structure. You couldn't ask for better fracturing of the concrete.

7:59 a.m. Precisely one minute before the shot, twelve long blasts from the air horn of a fire engine provide a final warning to anyone who may still be inside or nearby the doomed building. A thousand index fingers are poised above a thousand camera shutter buttons. Still, silence.

8:00 a.m. Down in the VIP tent, a child selected for the honor in a contest presses the button that connects the firing circuit, sending voltage

Have Dynamite, Will Travel

The building construction industry in the United States and Canada is huge, with tens of thousands of companies. The building demolition industry is, by comparison, tiny. Within that small community of demolition contractors there are less than a dozen companies specializing in "implosion" demolition.

Dykon is one of the biggest of these. Like the rest of the demolition industry, it is a small family business. Jim Redyke formed the company after working for another family demolition business for several years. Jim formed Dykon in 1974, and the fledgling company took on all sorts of construction blasting projects in addition to the demolition work.

Dykon's first blasting project was a pair of steel buildings in Atlanta, Georgia, in August 1974. The company went on to drop a 280ft concrete smoke stack in Tulsa, Oklahoma, a hospital complex back in Georgia, and a six-story concrete structure in Oklahoma. After that, the jobs flowed in steadily, and Dykon blew up (or down) smoke stacks (including one smoke stack drop that was filmed for the movie, *Hooper*), bridge piers, office buildings, a twelve-story hotel, a roller coaster (filmed for another Burt Reynolds film, *Smokey and the Bandit*). Within just a few years, Jim's little company had killed off dozens of structures of all shapes and sizes. On July 6, 1981, Dykon dropped a 900ft multiflue smoke stack at Johannesburg, South Africa, the world's tallest stack to be demolished with explosives. Dykon's current count is over 200 buildings and major structures. Ah, memories, memories!

---

to all the electrical blasting caps in all the holes throughout the structure. Down on the ground level, in the old lobby, the charges fire instantaneously with a sharp, hard, startling, BANG that echoes from the buildings surrounding the site as small puffs of dust squirt from under the fabric and chain link fencing.

Two seconds later, another BANG, this one much milder, as the columns on the second and third stories fire. Another set of small dust clouds reveal the location of the blasts, but nothing else happens.

After another brief interval, another set of charges fire. No chunks of concrete fly through space, no dramatic eruptions of material…but the front of the structure begins to slide toward the ground.

BANG, another set of charges fire. Now the elevator house on top of the building starts to lean. The basic structure remains essentially vertical, but the front of the building is shattered. It tears itself apart, progressively, just as Jim intended. The rear of the structure, without any explosives and reinforced by massive cables, provides a hinge for the collapse, anchoring the back of the building and forcing the decaying building to fall into what had once been a handsome and elegant entry.

The Pacific Palisades building took two years to build and millions of dollars to construct, but it only takes $200 worth of dynamite and ten seconds to drop it to the ground. The roof disappears into a massive cloud of dust, right where Jim wanted it to go. The long, rippling roar of the dying building echoes for a few seconds against the surviving buildings of Vancouver's skyline. That roar is replaced by another, this time from the audience who hoot and cheer and yell. It was, indeed, a fine performance.

Within minutes the crowds disperse, the film crews collect their cameras, and Jim, Corry, and Stan gather to inspect the damage and to congratulate each other. A news video crew interviews Jim, who looks happy and relieved.

Within an hour a little Case 580 is already using a hydraulic breaker to attack some of the rubble. Before the crowd has even fully dispersed, Pacific has two huge excavators—a massive, million-dollar John Deere 992 and an equally giant Hitachi—working on the rubble. The sooner the site is cleared the sooner construction can begin on the new building.

By nightfall, Jim and the rest of the Dykon crew—their role in the play completed—were on an airplane headed back to home base in Tulsa, leaving Pacific Blasting to finish clearing the site. They didn't get much time to recover from all the excitement, though, because of a bridge in Montana that needed to be brought down and a railroad depot "head-house" shot in Houston, Texas, to plan. As the Dykon company's unofficial motto says, "Have dynamite—will travel!"

Zero-plus-two seconds: About two seconds after the button was pushed, the big charges at ground level have all fired, notching the structure and beginning the failure sequence. It will certainly fall now, but nobody (except a few hundred blood-thirsty spectators, perhaps) wants it to fall over intact. A second set of charges has just detonated along the right side of the structure, along with those in some interior columns. While the building is still apparently intact, the lower floors already show evidence of structural failure.

Zero-plus-three seconds: Only one second later, the whole right side of the building collapsed and the elevator support structure began to lean crazily, while the back wall remained intact. The falling elements of the structure will pull the back wall away from the adjacent building, only twenty feet from the back wall. Charges continue to fire inside the building even though the firing circuit has probably been cut in dozens of places.

Zero-plus-four seconds: Shattered, but still held somewhat together by rebar and inertia, the Pacific Palisades starts toward the ground.

Zero-plus-five seconds: With the front of the structure tugging at the back wall, the destruction of the building is about half complete.

Zero-plus six-seconds: The last of the Pacific Palisades Hotel crumbles down to the ground.

Zero-plus-seven seconds: A dust cloud obscures the pile of rubble where a fine, modern building stood just a few moments before. Within a year or so a new apartment complex will occupy the site.

NEXT PAGES
That pile of concrete rubble was a twenty-two-story building just an hour ago. Two hundred pounds of dynamite, artfully placed, dropped the structure in the heart of Vancouver, British Columbia, and the excavators are already busy clearing the debris.

Index